ENGINEER DIVING OPERATIONS

U.S. Army Engineer School

Fredonia Books
Amsterdam, The Netherlands

Engineer Diving Operations

by
U.S. Army Engineer School

ISBN: 1-4101-0878-3

Copyright © 2005 by Fredonia Books

Reprinted from the 1992 edition

Fredonia Books
Amsterdam, The Netherlands
http://www.fredoniabooks.com

All rights reserved, including the right to reproduce
this book, or portions thereof, in any form.

ENGINEER DIVING OPERATIONS

TABLE OF CONTENTS

	Page
PREFACE	iii
CHAPTER 1. EMPLOYMENT OF ENGINEER DIVERS	1-1
Command and Control	1-2
Engineer Diving Support Priorities	1-3
Diving Support Request Procedures	1-3
CHAPTER 2. ENGINEER DIVING ORGANIZATIONS	2-1
Control and Support Diving Detachment	2-1
Lightweight Diving Team	2-3
CHAPTER 3. ENGINEER DIVING MISSIONS	3-1
Port	3-1
Clearance	3-3
Ship Husbandry	3-4
Physical Security	3-5
Logistics Over the Shore Operations	3-6
Offshore Petroleum Distribution Systems	3-6
River Crossing	3-7
CHAPTER 4. CONSIDERATIONS	4-1
Environment	4-1
Manning	4-3
Modes of Diving	4-3
Equipment	4-4
Medical Support	4-4
Safety	4-4
APPENDIX A. ENGINEER DIVING FORCE COMPOSITION TOE 05-530LA00 - C & S DIVING DETACHMENT	A-1
APPENDIX B. ENGINEER DIVING FORCE COMPOSITION TOE 05-530LC00 - LW DIVING TEAM	B-1
APPENDIX C. MINIMUM STAFFING LEVELS FOR VARIOUS TYPES OF AIR DIVING	C-1

	Page
GLOSSARY	Glossary-1
REFERENCES	References-1

PREFACE

This field manual (FM) provides a doctrinal basis for planning and using engineer divers in the theater of operations (TO). It describes responsibilities, relationships, procedures, capabilities, constraints, and planning considerations in the conduct of engineer underwater operations throughout the TO.

Its primary purpose is to integrate engineer underwater operations into the overall sustainment and mobility engineering structure. The doctrine presented is applicable to low-intensity conflict (LIC), combined, joint, and contingency operations.

This manual was designed for all commanders and planning staffs who require engineer diving assistance or those required to give engineer diving assistance.

The proponent for this publication is the United States Army Engineer School (USAES). Submit changes for improvement on Department of the Army (DA) Form 2028 (Recommended Changes to Publications and Blank Forms) to Commandant, US Army Engineer School, ATTN: ATSE-CDM-S, Fort Leonard Wood, Missouri 65473-6620.

Unless this publication states otherwise, masculine nouns and pronouns do not refer exclusively to men.

CHAPTER 1

EMPLOYMENT OF ENGINEER DIVERS

Engineer divers support all specialized underwater missions in the TO. The primary mission of engineer diving operations is to support sustainment engineering in the communications zone (COMMZ), providing a means for movement of logistics from port harbors and beach fronts to forward- and rear-area units. The secondary mission is to support maneuver units during water-crossing operations in the forward-battle area, providing maneuver units the capability to cross wet-gap obstacles while minimizing losses. Figure 1-1 is an example of engineer diver deployment in the theater.

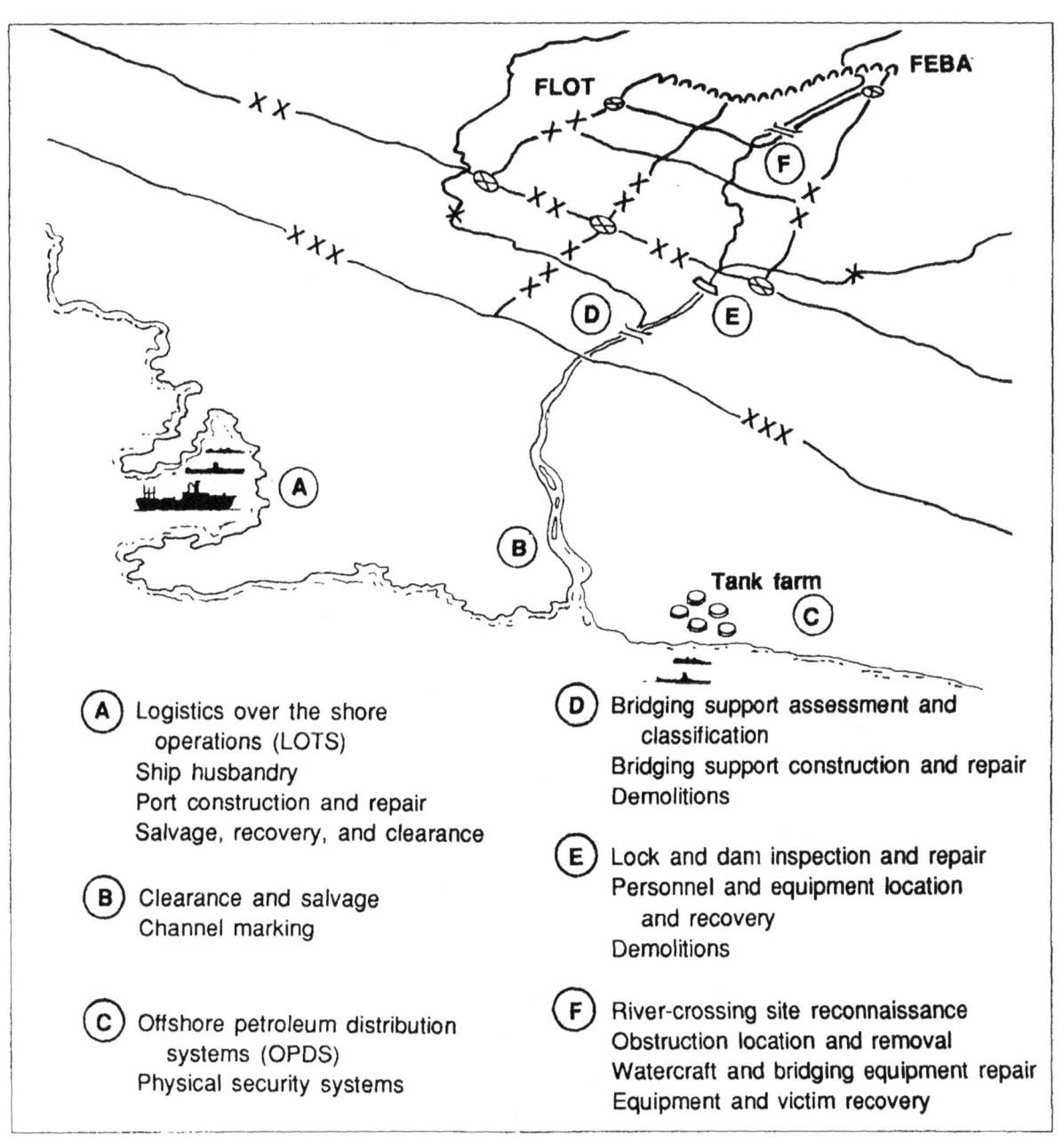

Figure 1-1. Engineer diver employment in the theater

COMMAND AND CONTROL

Engineers at the theater army headquarters (TAHQ) engineer command (ENCOM) formulate the plans and determine requirements for port facilities (location, capacity, wharfage, and storage). The theater Army (TA) is responsible for port operations and liaison with the US Navy, US Coast Guard, and other military and authorized civilian agencies from the US and allied countries. General responsibilities of the theater commander, TA commander, and the commander of the Theater Army Area Command (TAACOM) are stated in FM 100-16.

Theater construction and repair tasks that cross service boundaries and require divers will be managed by the regional wartime construction manager (RWCM). ENCOMs perform as the RWCM and provide command and control to the TA engineer force. The ENCOM is the echelons above corps (EAC) engineer headquarters responsible for constructing, maintaining, and repairing the theater sustainment base. When tasked, responsibilities include providing support to other allied military forces in joint or combined TO. The number and type of engineer units in the ENCOM depend on the size of the sustainment base, availability of host-nation support, and perceived threat to the rear area.

Engineer diving units are divided into two distinct organizations. The control and support (C&S) diving detachment provides command, control, and support of diving operations; the lightweight (LW) diving team executes most of the underwater work. The C&S diving detachment is assigned to ENCOM headquarters and may control up to six LW teams. The LW teams may be attached to units requiring prolonged diving support. Chapter 2 describes diving organizations in detail.

Figure 1-2 shows a typical ENCOM and TA interface for a theater Army engineer force.

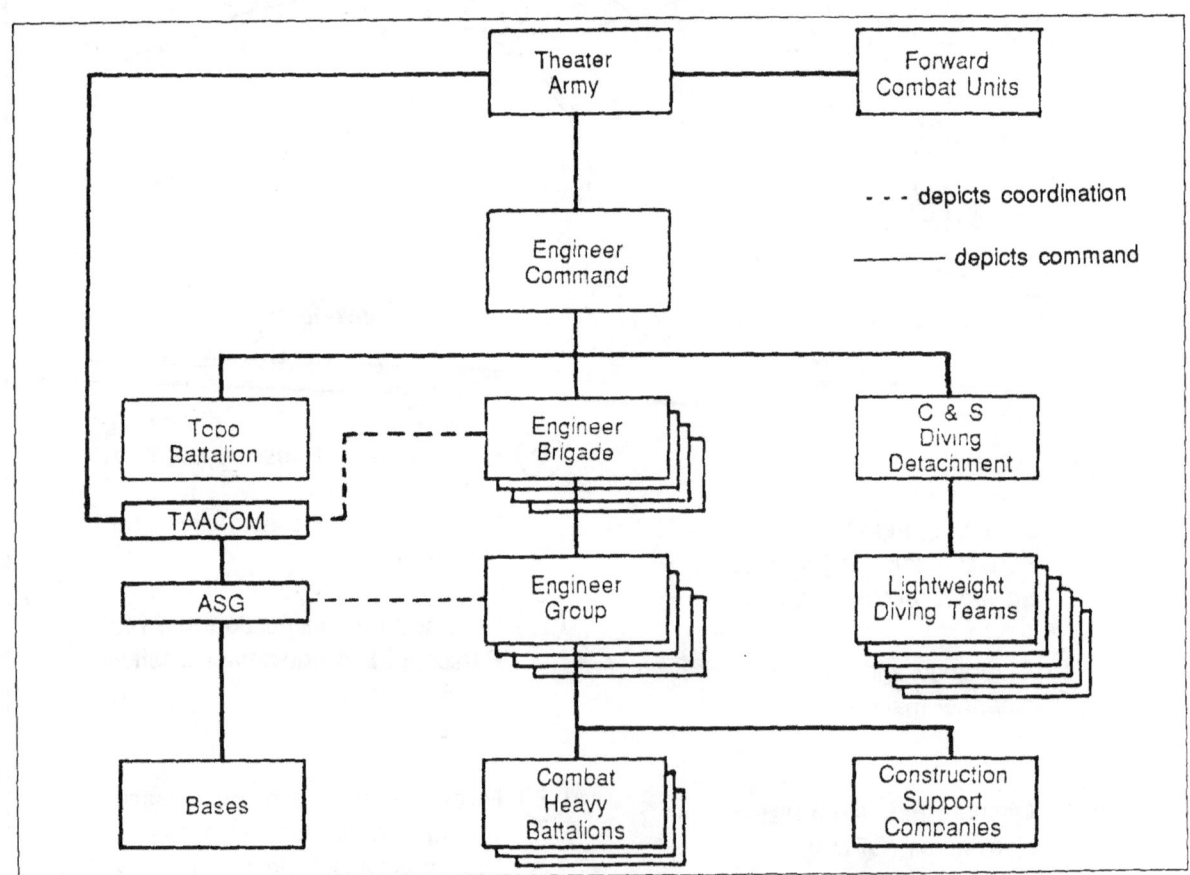

Figure 1-2. Engineer command structure and interface with theater Army

ENGINEER DIVING SUPPORT PRIORITIES

Engineer diving expertise is required throughout the theater. The ENCOM commander allocates assets in the COMMZ and combat zone (CZ) according to priorities established by the theater commander. Since there are only a limited number of divers, the ENCOM commander may choose to allocate diving assets only to the most critical mission sites.

The majority of underwater work performed by engineer divers requires the use of surface-supplied diving equipment. This requires time to move and set up before diving operations can begin. Once on site, a diving section may need several hours to prepare before deploying a surface-supplied diver into the water. It is critical to include planners from the C&S detachment during early planning stages of an operation to ensure successful diving missions.

The theater commander sets construction priorities and dictates policies which allocate construction assets and materials. The ENCOM commander, as RWCM, applies these policies in assigning diving assets throughout the theater.

Engineer diving tasks in the CZ usually support engineer mobility functions. In the COMMZ, the tasks usually center on sustainment operations such as port construction, harbor clearance, salvage, and ship husbandry. Divers also conduct immediate or emergency diving operations to help save lives or reduce equipment loss plus support interservice recovery operations.

DIVING SUPPORT REQUEST PROCEDURES

After completing mission and situation analyses, the ENCOM commander attaches divers to the appropriate organizational level. Figure 1-3 illustrates request channels for engineer divers.

• COMMZ. If an area support group (ASG) requires diving assets for underwater missions, the requests are forwarded to the theater command. Requests must include mission details and estimated time for work completion. Approved requests will be tasked to the ENCOM by the theater commander, who assigns diving priorities. For short-term missions, diving assets are assigned in direct support through command channels to the ASG. For long-term or complex missions, divers are normally attached to a company- or battalion-size unit. For example, if an ASG port construction company needs diving assets for port repair, the ENCOM commander will attach diving teams to the construction company.

• CZ. CZ tasks may involve survey of river-crossing sites, location or removal of underwater obstacles, repair of watercraft, and recovery of lost equipment. Approved requests will be tasked to the ENCOM by the theater commander. The ENCOM commander will attach divers in direct support to the appropriate organizational level. Chapter 3 discusses engineer tasks and support relationships.

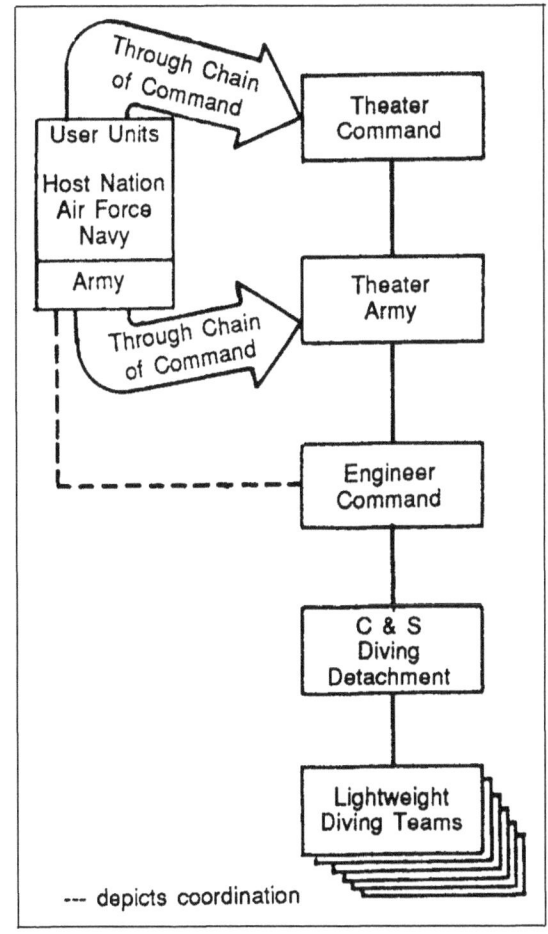

Figure 1-3. Request channels for engineer divers

- Air Force. Separate engineer divers support Air Force immediate recovery operations for downed aircraft in ports or water locations near the shore. The theater Air Force manager for these operations is the survival recovery center (SRC). The SRC coordinates closely with the ENCOM. Air Force requests for immediate recovery operations go directly to the ENCOM, which responds in accordance with (IAW) theater mission priorities. Immediate recovery operations are usually assigned to divers as an on-order, direct-support mission.

- Navy.

    - Engineer divers may support Navy operational commitments for construction, salvage, or watercraft maintenance. Navy maintenance organizations request diving support from the Army water terminal commander located in the port facility. If divers are not currently attached to the terminal organization, the Army water terminal commander forwards the request through command channels to the ENCOM, detailing the need for divers to support naval operations. If approved, the ENCOM will task Army divers to support the Navy mission.

    - If divers are on site supporting Army terminal operations, the Army water terminal commander may task the Army diving teams to support a specific naval maintenance unit. This is based on work priorities and higher command guidance.

- Host nation.

    - Support to the host nation is common during port construction, repair, and clearance. Requests for engineer divers are approved at theater Army and tasked to the engineer or transportation command to which divers are assigned. The ENCOM commander will attach the diving teams to the appropriate command to support the mission.

    - Divers may also support host-nation immediate recovery operations for civilian aircraft or equipment downed in ports or bodies of water near the shore. Civilian authorities request divers directly from the nearest engineer battalion, brigade, or ASG. These units forward requests to the ENCOM for approval. The assignment of diving support is IAW theater guidance and work load. Immediate recovery operations are usually assigned to divers as an on-order, direct-support mission.

CHAPTER 2

ENGINEER DIVING ORGANIZATIONS

Engineer diving units are relatively small, specialized organizations. Each detachment or team has specific duties and responsibilities but is flexible and improvises to support the theater in most situations. Diving units are subordinate elements of the theater's ENCOM. They normally provide general support to the theater. When required, they also provide direct support to commanders below theater level.

Diving units are divided into two tables of organization and equipment (TOE). The C&S diving detachment, TOE 05-530LA00, is assigned to EAC and provides command and support to LW teams. The LW diving team, TOE 05-530LC00, is assigned to the C&S diving detachment and can be attached to supported units during the execution of diving missions. Although assigned to EAC, the C&S augments LW teams during missions requiring additional diving support.

CONTROL AND SUPPORT DIVING DETACHMENT

Each C&S detachment can support up to six LW teams within the theater. The C&S detachment has an organic scuba team used for inspection and survey. The C&S detachment monitors the current and projected work load of its organic scuba team and assigned LW teams.

The C&S detachment has 13 soldiers and sufficient equipment to provide specialized support when required by the LW teams. The detachment has a command, control, and operations section; a supply/prescribed load list (PLL) section; an equipment maintenance section; and manpower for a 7-man scuba team. The team can dive independently or augment a LW team to provide manpower needed for deep-sea diving. The C&S scuba section is limited by its equipment to performing only inspection, survey, and damage assessment missions.

The C&S detachment provides the following support to EAC commands:

• Staff assets for theater diving integration and planning.

• Command and support of all assigned diving assets.

• Interservice liaison.

• Theater command diving expertise.

• A scuba section for damage assessment and premission assessment (site reconnaissance).

The C&S detachment provides assigned LW teams with the following specialized support:

• Mission analysis and planning.

• Special life-support diving equipment.

• Divers, equipment, and support personnel when augmenting into a deep-sea team.

• Repair parts and supplies for engineer diving life-support equipment.

• Diving equipment maintenance.

• Recompression chamber treatment augmentation.

• Diving medical support.

• Requalification and diver training.

When the C&S detachment augments the LW team to perform deep-sea missions, scuba inspection capabilities are lost. Initial survey reports and availability of diving assets influence routine work priorities. See Chapter 4 for normal working depths.

Figure 2-1, page 2-2, illustrates a typical engineer diving organization and assigned LW teams. Figure 2-2, page 2-2, shows the organization of a LW team.

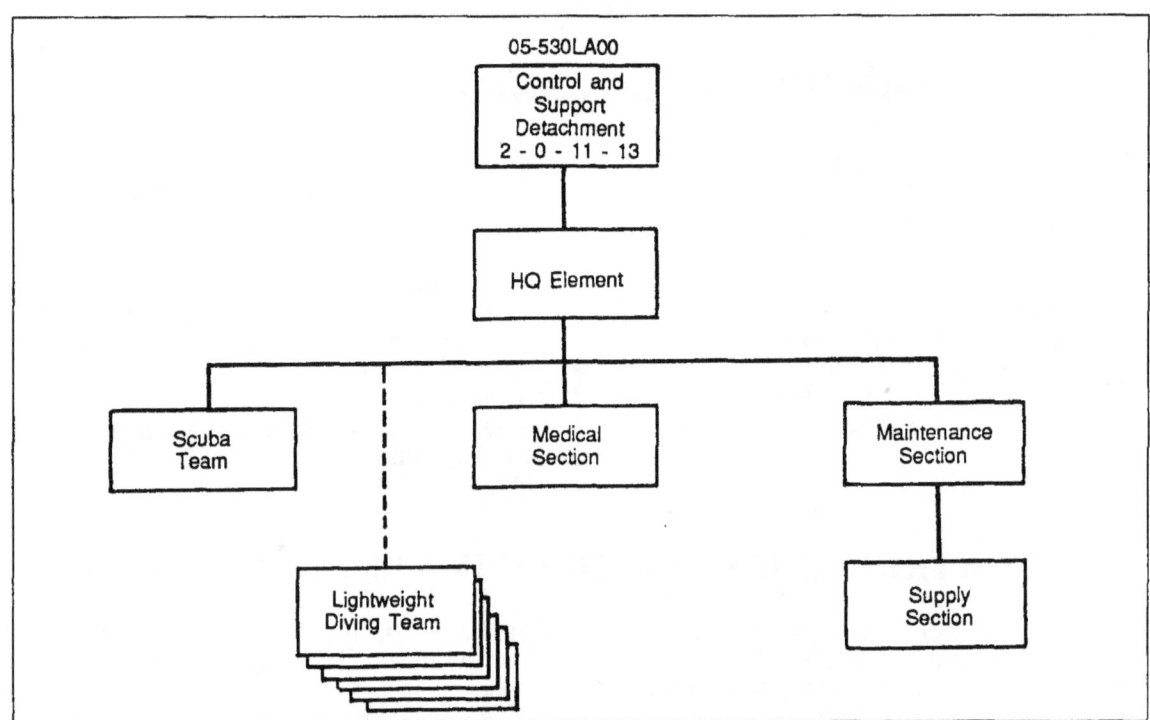

Figure 2-1. Organization of C&S diving detachment with LW diving teams assigned

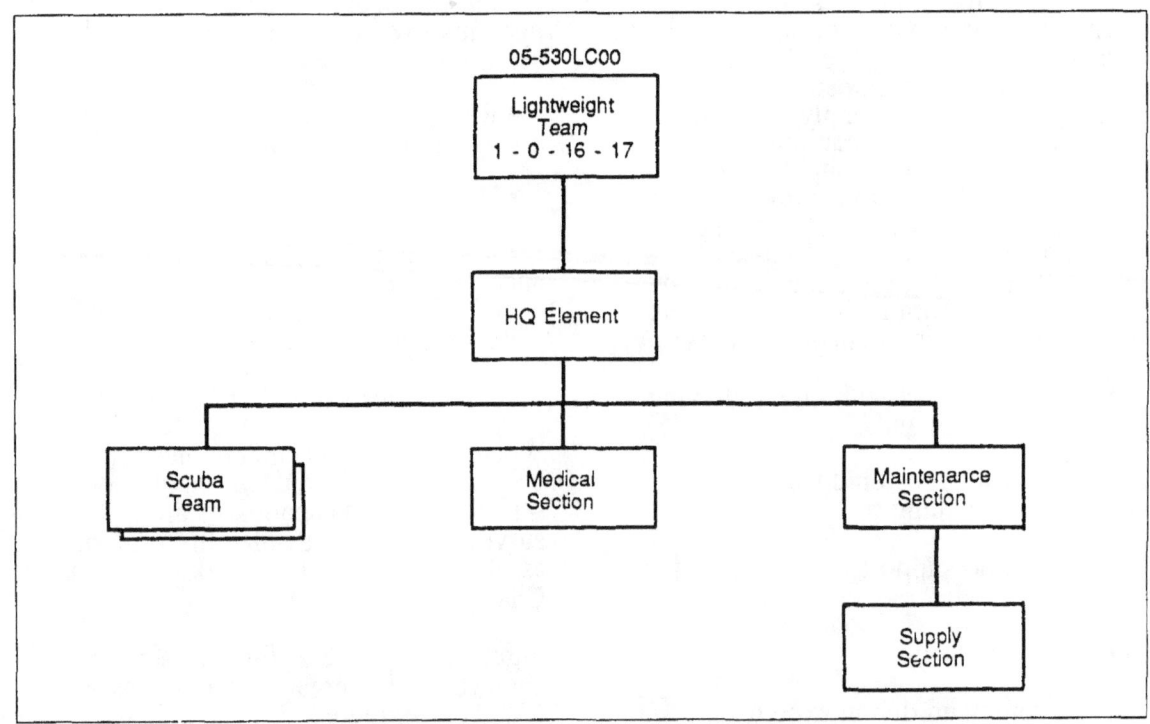

Figure 2-2. Organization of LW diving team

Appendix A lists C&S detachment manpower and special equipment allotted in the base TOE. C&S detachment personnel have the following responsibilities:

• Detachment commander--CPT. (Must be a qualified diver.) Responsible for all diving operations within the theater. Commands and supervises detachment and assigned LW teams throughout the theater. Performs as the diving officer during deep-sea diving operations.

• Operations officer--1LT. (Must be a qualified diver.) Coordinates all diving missions. Assigns diving missions to teams and sections. Plans and schedules required training for the C&S detachment and assigned teams. Performs as executive officer and supply officer for C&S and LW teams.

• Senior diving supervisor--MSG. (Must be a qualified diver.) Senior diving supervisor in the theater and responsible for the safe conduct of all diving operations within the theater. Performs duties as the detachment's master diver. Supervises deep-sea and demolition diving missions. Assists the commander and operations officer during planning, scheduling, and executing training and operational missions. Coordinates medical treatment with the theater Army diving medical physician. Provides diving expertise to staff planners and assigned LW teams.

• Diver--SSG. (Must be a qualified first-class diver.) Supervises diving teams during scuba and surface-supplied diving operations. Works closely with the master diver and diving officer during preparation of the operations order; plans the dive step by step. Determines equipment requirements and assigns divers to specific tasks. Supervises maintenance of all diving equipment and associated diving-support equipment through intermediate-level maintenance.

• Supply sergeant--SSG. Supervises the PLL clerk in support of specialized PLL required for the detachment and assigned teams. Performs as the supply sergeant for the detachment and assigned LW teams, maintaining diving supplies and repair parts through intermediate direct support maintenance (IDSM) and intermediate general support maintenance (IGSM) levels. Coordinates depot-level repair for diving life-support equipment.

• Diver--SGT. Performs as a diver on the detachment scuba inspection section or when augmenting the LW teams. Performs maintenance and repair functions on diving life-support equipment for the detachment and assigned teams through the intermediate levels of maintenance.

• PLL clerk--SGT. Performs as PLL clerk in support of all diving life-support equipment and is supervised by the detachment supply sergeant. Maintains diving equipment specific PLL in support of the detachment and assigned LW teams.

• Emergency treatment noncommissioned officer (ETNCO)--SGT. (Must be a qualified diver.) Performs as diving medical technician inside the recompression chamber. Assists the diving supervisor in diagnosing and treating diving-related illnesses and injuries. Performs maintenance on the recompression chamber facility. Coordinates training and medical supplies with the theater Army diving medical physician. Assists the commander and operations officer in planning and scheduling training requirements for ETNCOs in the assigned LW teams.

• Diver--SPC. Performs as diver on the detachment scuba inspection section or when augmenting LW teams. Under supervision, performs maintenance on all diving equipment and associated life-support equipment.

LIGHTWEIGHT DIVING TEAM

The LW team is assigned to a C&S detachment which provides mission tasking and specialized diving support. LW teams support engineer groups responsible for key port facilities or major logistics over the shore operations (LOTS) mission support. They may be attached to units requiring extensive diver support for ship husbandry, underwater pipeline maintenance, port construction, and other missions requiring underwater maintainance of waterborne lines of communication (LOC). Typically, supported units are engineer port construction and bridging companies, transportation floating craft general maintenance and boat companies, and quartermaster marine pipeline terminal companies.

The LW teams provide underwater support for diving missions including--

• Support of bridging and other water-crossing site surveys.

• Search, salvage, and recovery of submerged tools, equipment, weapons, and vehicles.

• Removal of submerged obstacles from navigable waterways using underwater demolitions or underwater cutting and welding techniques.

• Inspection and repair of damaged bridges, piers, docks, and related structures.

• Security inspections of critical bridges and other structures against sabotage.

• Search for and recovery of water casualties.

• Construction of security screen for critical bridges, piers, docks, wharves, quays, and associated port facilities.

• Inspection and repair of watercraft.

• Inspection of US support ships to prevent sabotage.

The LW teams also provide special support for the following areas which were discussed in Chapter 1:

• Theater operations.

• Air Force operations.

• Navy operations.

• Host-nation support.

The LW team has 17 soldiers and sufficient equipment to deploy one LW team or two 7-man scuba teams as shown in the diving organization diagram (Figure 2-1, page 2-2). The LW team performs diving missions to water depths of 190 feet. The normal work shift for the LW team is 12 hours. Diving missions that require continuous 24-hour operations, or working environments requiring total diver enclosure for protection, are considered deep-sea missions. The LW team can support deep-sea missions when augmented by additional personnel and equipment from the C&S detachment.

Air compressors and high-pressure air flasks, located on the surface, provide breathing air for both the LW and deep-sea teams. Self-contained, man-portable cylinders provide breathing air for scuba teams.

Appendix B lists the LW team's manpower and special equipment identified in the base TOE. LW teams have the following individual responsibilities and capabilities:

• Diving team leader/diving officer--1LT. (Must be a qualified diver.) Coordinates and plans diving missions. Responsible for operations and mission accomplishment of the diving unit. Assists and performs as backup to the diving supervisor and master diver during decompression dives or recompression chamber operations. Performs equivalent duties of a platoon leader.

• Senior diving supervisor--SFC. (Must be a qualified master diver.) Supervises surface-supplied diving missions and recompression chamber operations. Assists the team leader in planning, scheduling, and executing training and mission requirements. Provides diving expertise to staff planners.

• Diver--SSG. (Must be a qualified first-class diver.) Works closely with the master diver and diving officer during preparation of the operations order. Supervises scuba and surface-supplied diving operations. Is responsible for planning the dive, selecting and setting up the diving equipment, and briefing the divers. Supervises maintenance of all diving equipment and associated diving-support equipment through intermediate levels of maintenance. A diving supervisor must be present during all dives.

• Diver--SGT. Performs as a diver during surface-supplied and scuba diving missions. Performs unit maintenance and repair on diving equipment and associated life-support equipment.

• ETNCO--SGT. (Must be a qualified diver.) Performs as the diving medical technician inside the recompression chamber. Assists the diving supervisor in diagnosing and treating personnel for diving-related illnesses and injuries. Maintains the recompression chamber facility. Coordinates diving medical treatments and training with the C&S detachment ETNCO.

- Diver--SPC. Performs as a diver during surface-supplied and scuba diving missions. Under supervision, performs unit maintenance of all diving equipment and associated life-support equipment.

CHAPTER 3

ENGINEER DIVING MISSIONS

Engineer divers help keep the waterborne LOC open. They also support the forward movement of troops and equipment. This support ranges from a 5-man scuba team to deploying a 19-man, deep-sea team using surface-air-supplied breathing equipment. The scuba teams perform inspections, surveys, searches, recoveries, and light work. The deep-sea teams perform extensive diving operations during heavy salvage, construction, or harbor clearance missions.

Seven major essential missions are identified for engineer divers. The missions include--

- Port.
  - Planning and inspection.
  - Construction.
  - Repair.
- Clearance.
  - Salvage.
- Ship husbandry.
  - In-water hull inspections.
  - In-water maintenance.
  - Damage control and repair.
- Physical security.
  - Physical security systems.
  - Security swims.
- LOTS.
  - Hydrographic surveys.
  - Salvage and mooring systems.
  - Petroleum pipeline.
- Offshore petroleum distribution systems (OPDS).
  - Permanently installed submarine pipeline.
  - Single anchor leg mooring system (SALMS).
  - System repair and maintenance.
- River crossing.
  - Survey-crossing site.
  - Obstacle location and removal.
  - Equipment recovery.
  - Bridge inspection and repair.
  - Retrograde operations.
  - Personnel recovery.

PORT

PLANNING AND INSPECTION

Preliminary and detailed construction planning is an overall guide for construction activities and is accomplished prior to beginning construction work. Planning should include formulating a strategy for clearing sunken vessels and obstructions from within the port area. The ENCOM headquarters should include a qualified planner from the C&S diving detachment to identify diving requirements and to ensure proper allocation of diving assets. The C&S detachment assists in the development of a construction plan and provides a scuba inspection team for initial on-site surveys. After completing initial inspections, the C&S detachment will designate the appropriate diving team most capable of performing the mission. The C&S detachment augments the LW team with personnel and equipment for missions requiring extensive diving assets such as major salvage, construction, and harbor clearance. Planning and initial inspections include--

- Conducting initial on-site, underwater surveys to determine the possibility of restoring the port facilities (piers, quays, wharves, dry-dock facilities, marine railway systems, and other port structures) to an operational status.

- Inspecting damaged and sunken vessels and other obstructions in the port to determine requirements for salvage or removal.

- Assisting in development of a salvage strategy for clearing the port area and ship channels.

- Assisting in development of time estimates for salvage and clearance.

- Assessing underwater damage of existing pier facilities.

- Estimating time for underwater construction.

An underwater assessment survey will provide the Army water-terminal commander with a report of existing conditions of underwater port facility structures. A port-bottom profile depicting water depths and obstruction locations will be included in the report. Information provided will assist the area engineer and port construction units in determining the scope of construction required for port repair. It will also assist them in developing a port repair plan and time estimate. A detailed report will include--

- Recommendations for restoration.

- Location and condition of sunken vessels or other obstructions.

- Water depths of ship channels within the port.

- Recommendations for vessel or obstacle removal.

- Location of underwater mines and munitions.

To ensure timely procurement of needed materials, divers must make a detailed underwater survey and assist in developing the bill of materials for repair missions.

NOTE: Engineer divers can clear mined areas from the surface through the use of sympathetic detonation with demolitions. The divers can also mark suspected mined areas or identify mines for removal by qualified Navy explosive ordnance disposal (EOD) teams. The Army does not have EOD-trained diving teams.

CONSTRUCTION

The construction of new ports and facilities is a major undertaking which usually requires extensive use of divers. Divers can provide valuable information during initial site selection and survey.

Hydrographic surveys of the proposed area are conducted to determine water depths, sea-bottom contours, and location of ship channels and underwater obstacles. Hydrographic surveys are covered in the Logistics Over the Shore Operations section of this chapter.

The technique for assembling and installing underwater components is similar to the method used on the surface. Detailed port construction techniques are outlined in FM 5-480. Underwater construction and repair techniques can be found in NAVFAC P-990.

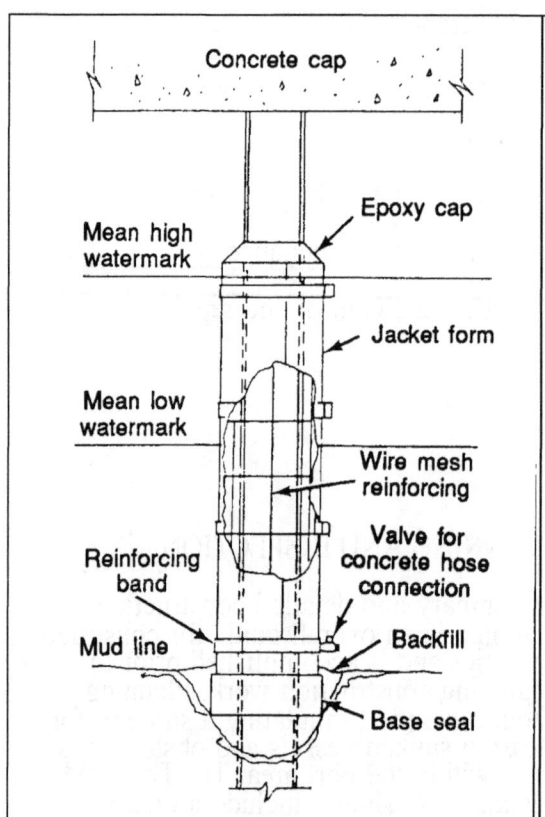

Figure 3-1. Concrete protective jacket around a timber pile

REPAIR

The repair method depends on the original construction material, type of repair material, and degree of repair desired.

Divers can perform underwater repair of bearing piles, fender and dolphin systems, and support walls. Underwater structures must be thoroughly cleaned before inspecting and repairing. Inspection and repair of these structures require specialized equipment. Repairs can be as simple as filling minor cracks with special epoxy; installing a concrete protective support jacket; or replacing wooden, steel, and concrete supporting structures and hardware. Figure 3-1 illustrates a damaged woodpile repaired with a concrete protective jacket. Repairs can be as extensive as major rehabilitation and replacement of the underwater structure supports.

Concrete is often used to repair port structures. Underwater concrete placement techniques are basically the same as surface applications; however, the requirement for the diver to wear awkward diving equipment and to work in a zero-visibility environment greatly increases the difficulty of the operation.

Steel is normally used to repair wooden structures such as bearing piles, piers, and fender systems. The repair of steel structures is complex and normally requires thorough cleaning and underwater welding. Special equipment designed for underwater work must be used, and strict safety rules must be followed. High-voltage electricity passes through the water to the welder. This increases the risk of electrical shock.

CLEARANCE

Clearance operations are undertaken to neutralize all obstacles blocking the shipping channels in ports, docking facilities, mooring sites, marine railways, dry-dock facilities, lock and dam structures, and other navigable waterways. Clearance consists of locating, marking, surveying, and removing underwater obstructions. The operations include removal of natural (underwater rock formations) or man-made obstacles, battle debris, or enemy-emplaced objects intended to prevent the use of navigable waterways or port facilities.

Various methods are used for removing obstructions. These methods include using lifting bags and other equipment from underwater salvage tool kits, demolition charges, cranes, or underwater cutting equipment. Additional lifting force is usually obtained from various items such as empty 55-gallon drums or fuel container bags commonly found in a port facility.

Demolitions provide an efficient method for removing underwater obstacles in the port area. Most explosives are designed for underwater use; however, their effectiveness is sometimes degraded due to the change in environment from air to water. For example, shaped charges require a low-density material, such as air, directly underneath the main charge. This low-density material is changed upon entry into the water.

Special precautions are required when employing demolitions underwater. Electric firing systems should be used whenever possible to control the charge detonation, thus increasing diver safety. Safe distances must be extended because of increased distance and density of the shock impact resulting from water pressure. Charges detonated near any vessel or personnel in the water can cause substantial damage or injury.

Underwater cutting operations are usually required to reduce an obstacle to manageable size for removal. Special underwater cutting and welding sets are available. Hydraulic, pneumatic, and special hand-tools increase work efficiency.

NOTE: Diving teams normally require the supported unit to provide a welding machine for this mission (a 400-ampere power source is required).

SALVAGE

Major salvage operations include the clearance and removal of sunken vessels, equipment, supplies, or other materials from port channels, berthing and docking facilities, mooring sites, lakes, lock and dam facilities, and other navigable waterways. The diver's ability to salvage vessels or other equipment depends on the type, size, and location of the object and time available for the salvage effort. Methods of salvage range from simple hole-patching and dewatering to completely dismantling a vessel into sections for removal. Beached vessels resting on the bottom with the superstructure above the mean low watermark are salvaged by patching exterior holes and dewatering the hull. The vessels can then be towed to another location for repair by qualified personnel.

Sunken vessels with the superstructure below the mean low watermark require more extensive salvage efforts. Divers must make the entire vessel watertight, which usually means penetration dives into the vessel interior for inspection and repair work. The vessel is then lifted by dewatering, attaching underwater lifting devices to the hull, or lifting with surface-support cranes. Sometimes a combination of these techniques is necessary.

Unsalvageable vessels and other equipment can be marked and left in place, sectioned and removed, or flattened with demolitions. Sectioning means cutting into manageable pieces and then removing to designated locations.

NOTE: Removing large sections of steel may require a surface crane or winching machine from the supported unit. Flattening includes using demolitions to remove the superstructure and crushing the hull into the port bottom.

SHIP HUSBANDRY

Ship husbandry is the in-water inspection, maintenance, and repair of vessels. Troops, equipment, and supplies are transported using Army vessels. Army divers are tasked to provide maintenance assistance for these vessels. The ability of divers to perform ship husbandry depends on the following:

• Size and number of vessels requiring support.

• Number of divers available.

• Additional equipment and spare parts needed.

• Mission priority established by the Army water-terminal commander.

The C&S detachment commander can assist the Army water-terminal commander during coordination of ship-husbandry operations. The diving supervisor has overall control and responsibility for diver safety.

Special safety precautions for husbandry operations include--

• Direct coordination between the on-site diving supervisor and vessel master prior to the diver entering the water.

• Vessel shutdown and tag out of all systems that may endanger the working divers.

• Coordination with the harbor master to control vessel traffic in the vicinity of the diving operation.

IN-WATER HULL INSPECTIONS

In-water inspections of military vessels are performed to assess the condition of the underwater hull and appendages. The inspections cover all parts of the vessel below the waterline and are part of the scheduled maintenance or damage assessment. The inspection provides the vessel master with information necessary to determine the condition of the vessel.

Vessel appendages include all zinc anodes, heat exchangers, sonar domes, depth finders, and any exterior-mounted system. In-water hull inspections provide the vessel master information on the following vessel components and appendages:

• Hull. Damage assessment and identification of build up from marine organisms growing on the hull, plus condition of anti fouling paint surfaces.

• Propulsion and steering systems. Condition of shafts, screw propellers, and rudders and the serviceability of protective coatings, seals, and bearings.

• Vessel appendages. Determination of general condition and operational ability.

IN-WATER MAINTENANCE

In-water maintenance of military vessels is performed for scheduled maintenance or deficiency correction. In-water maintenance enables the Army water terminal commander to have immediate use of his watercraft. He can also keep the marine railway, dry dock, and other vessel maintenance facilities open for vessels requiring maintenance and repairs that divers cannot perform in water.

Divers provide in-water maintenance of the following military vessel systems:

• Propulsion and steering. Divers assist in repairing or replacing in-water components of the propulsion and steering systems. The supported unit must supply demolition and crane support, when required to aid in removing and positioning new components.

• Sea chest and heat exchanger. These appendages provide cooling to the various power plants on board the vessel. They are easily cleaned in the water using underwater hydraulic equipment and hand-tools. For more extensive repairs, divers can remove items for repair on the surface.

• General systems. Other maintenance includes the clearing of lines, ropes, or other debris from the propeller or the cleaning of any appendage located below the waterline.

DAMAGE CONTROL AND REPAIR

Damage control and repair provide immediate assistance to a vessel in distress. Repairs are temporary in their application and are meant to keep the vessel afloat until permanent repairs are made. Divers can provide assistance ranging from installing small damage control plugs to welding large patches. The vessel commander will direct repair in coordination with the on-site diving supervisor.

PHYSICAL SECURITY

Physical security operations include developing active and passive security systems to protect or provide early warning of impending danger to ports, channels, or pier facilities.

Divers can assist in placing and maintaining permanent physical security systems in port areas, upon fixed bridges, and in waterway lock and dam systems. Divers also perform security swims for waterborne vessels. The request for diving support must include the type of physical security system used.

PHYSICAL SECURITY SYSTEMS

Physical security systems are usually placed at harbor entrances, along the open areas of port facilities, and around bridge abutments. The systems may be passive or active and are designed to stop or detect vessels, underwater swimmers, or floating mines. These systems usually require diving support for installation and maintenance.

Passive security systems require introduction of obstacles or barriers that restrict the approaches and entrance to a harbor. Barriers across a harbor's access channel usually require constant maintenance and repair. Electronic security systems are designed to detect and, in some cases, deter attack by underwater swimmers. Divers place and secure the systems underwater after qualified personnel assemble the systems on the shore.

SECURITY SWIMS

Divers can perform physical security swims on the underwater portion of a vessel before it enters the port facility or while it is moored outside the secured perimeter. Although divers are capable of performing these inspections, they cannot remove any foreign explosive devices found during the inspection. The removal of these devices is the responsibility of underwater EOD teams. Periodic security swims are necessary on installed physical security systems to detect maintenance requirements and sabotage.

LOGISTICS OVER THE SHORE OPERATIONS

LOTS are the water-to-land transfer of supplies to support military operations. They are conducted over unimproved shorelines and through partially destroyed, fixed ports; shallow draft ports not accessible to deep-draft shipping; and fixed ports that are inadequate without using LOTS capabilities. Divers are an important asset during LOTS because of the large number of watercraft involved in the transfer of supplies. The scope of LOTS depends on geographical, tactical, and time considerations.

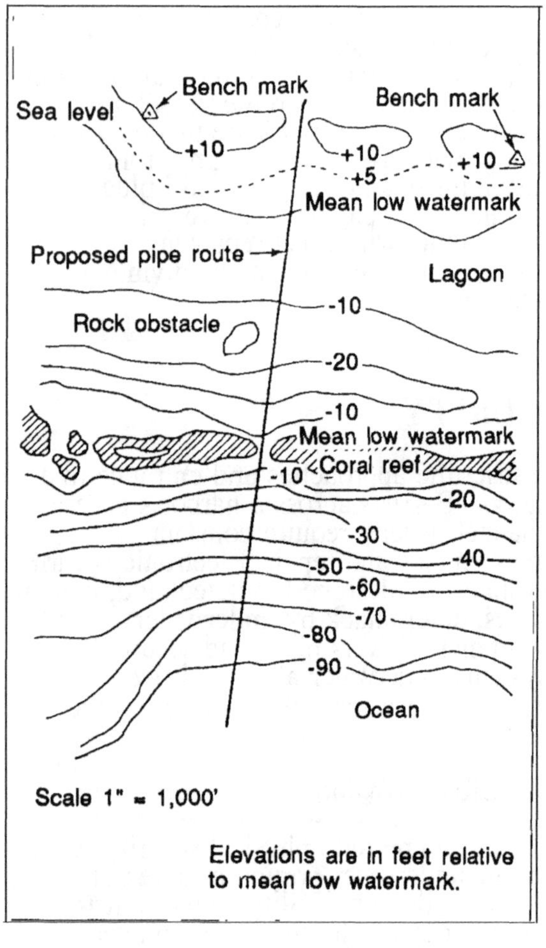

HYDROGRAPHIC SURVEYS

Hydrographic surveys provide the port or LOTS commander with a detailed chart depicting underwater bottom profiles of an operational shoreline or port area. This chart indicates bottom depth gradients, ship channels, and location and type of obstructions which may impede vessel traffic. Figure 3-2 illustrates a typical hydrographic survey for a proposed underwater water pipeline.

SALVAGE AND MOORING SYSTEMS

Unloading and transporting supplies at sea may result in the loss of supplies into the water. Divers can recover these supplies quickly and assure continued support to fielded units. They can also assist vessel crews by unfouling anchor lines or clearing debris caught in the propellers. In addition, divers can install and maintain offshore mooring systems to provide safe anchorage to cargo vessels, causeways, and landing craft supporting LOTS.

PETROLEUM PIPELINE

Divers provide underwater support during the installation of OPDS used in LOTS. Divers can perform the surveys necessary to determine pipeline positioning, assist in the actual pipe placement, and provide underwater pipeline inspection and maintenance.

Figure 3-2. Hydrographic survey

OFFSHORE PETROLEUM DISTRIBUTION SYSTEMS

Petroleum distribution systems are used extensively during fuel transfer operations. The transfer of fuel from tankers to the high watermark on shore is a Navy responsibility in joint area operations. However, the engineer port construction companies, engineer diving teams, and transportation watercraft groups play prominent roles in the preparation, installation, repair, and operation of the OPDS in Army theaters.

PERMANENTLY INSTALLED SUBMARINE PIPELINE

The construction of a permanently installed submarine pipeline is not expected during mobilization. However, systems already in place may require extensive repair and maintenance.

SINGLE ANCHOR LEG MOORING SYSTEM

The SALMS provides a semipermanent installation for bulk transfer of fuel directly from an offshore tanker to port storage. This system will be employed during OPDS operations, and divers may be required to support it by --

• Performing hydrographic surveys to determine beach gradient and underwater contour.

• Improving beach approaches.

• Clearing enemy-emplaced or natural obstacles from beach approaches.

• Supporting the installation of an OPDS.

• Connecting underwater pipeline components.

• Inspecting pipelines and their components.

• Performing maintenance on underwater pipeline components.

• Performing emergency repairs to damaged pipe sections.

SYSTEM REPAIR AND MAINTENANCE

The underwater components and mooring assemblies for all types of distribution systems require periodic maintenance support. Specific areas of repair and maintenance performed by divers are--

• Tanker hose discharge assemblies. These connecting hoses are of various types and require periodic replacement of gaskets and damaged sections. Control valves located at pipeline connections require periodic lubrication and seal replacement.

• Mooring systems. Mooring systems prevent ship movement during petroleum transfer operations. Maintenance includes periodic inspection and replacement of chain hardware connections and worn chain sections. Surface marking buoys require periodic cleaning and replacement.

• Pipelines. Permanently installed pipelines need periodic inspection and maintenance to ensure watertight integrity. Divers repair or replace pipe flange connections, damaged pipe sections, and concrete encasements. Divers conduct security swims along the length of the pipeline to verify pipeline integrity.

RIVER CROSSING

Divers are capable of providing support during river-crossing operations. Most missions are accomplished by separate scuba sections from the C&S detachment or LW teams.

SURVEY CROSSING SITE

Divers survey proposed river-crossing sites by performing bottom and underwater bank approach profiles. They locate, mark and, if necessary, remove underwater obstacles.

NOTE: Intelligence collection along enemy-controlled shores is not a function of engineer diving units described in this manual. It is performed by Special Operations Forces divers trained and equipped for unsecured area operations (FM 31-25).

OBSTACLE LOCATION AND REMOVAL

Divers assist in neutralizing underwater obstacles. They use sympathetic detonation to clear in-water munitions. This is accomplished by emplacing demolitions on or near underwater obstacles. Demolitions are always detonated from the surface. A clear lane is verified by dragging a cable or weighted line in the specified areas.

EQUIPMENT AND PERSONNEL RECOVERY

Divers assist in the recovery of sunken equipment and tools and provide water casualty search and recovery.

BRIDGE INSPECTION AND REPAIR

Divers perform in-water repair of float and fixed bridging. They also provide damage assessment and help determine bridge trafficability.

RETROGRADE OPERATIONS

Divers support retrograde operations by placing underwater demolition charges on bridge supports, anchorage systems, and salvageable equipment to prevent enemy use.

CHAPTER 4

CONSIDERATIONS

When planning, allocating, and executing a diving operation, careful consideration must be given to the following:

- Environment.
- Manning.
- Equipment.
- Medical support.
- Safety.

ENVIRONMENT

The mission, available divers, and weather help determine the type of diving and the equipment used. Surface-supplied diving provides the best safety for the diver and enhances the supervisor's ability to control and direct the divers underwater. Special equipment may be required to provide additional protection for the diver in extremely cold or polluted waters. Factors which influence the selection and deployment of diving teams include--

- Water temperature, depth, and pollutants.
- Current.
- Tides.
- Visibility.
- Bottom condition and type.
- Sea state and wave height.
- Air temperature.

Table 4-1. Diving limitations

Type of Equipment	Maximum Water Depth (ft)	Water Current (knots) (fps)	Duration Under-Water* (min)	NBC Protection	Environmental Protection	Salt-Water Temp (°F)
Deep-sea	190	2.5 / 4.2	40	None	Best	28
LW	190	2.5 / 4.2	40	None	Limited	28
Scuba	130	1.0 / 1.7	10	None	Limited	28

*Also limited by individual diver endurance and type of thermal protection worn.

If water currents exceed the maximum limits listed in Table 4-1, page 4-1, alternative methods should be considered. If employment of a diver is necessary, he must be afforded the highest margin of safety.

DECOMPRESSION

The time a diver can spend underwater is limited by physical considerations. Most of the work should be performed on the surface to minimize the amount of time a diver must spend underwater. The ability to perform work underwater is impaired by poor visibility, restricted movement (by diving equipment and bottom conditions), and limited time. Decompression requirements are a major concern to the diving team. Decompression obligations limit the amount of time a diver can remain on the bottom. As water depths increase, the amount of time a diver may safely spend underwater decreases. Dives are classified as either decompression or no-decompression dives.

When air is breathed under pressure, nitrogen from the air is absorbed in the tissues of the body. A diver's body absorbs and stores excess nitrogen whenever exposed to pressures found at water depths of 40 feet and deeper. The amount of nitrogen absorbed by the tissues increases with depth and time. The water temperature and the diver's physical condition and activity influence the amount of nitrogen stored in the tissues. During ascent, the pressure on the body is reduced, and the nitrogen is released from the tissues and is eliminated through normal respiration. It is essential to control the rate and delay ascent to allow the body sufficient time to process the nitrogen from the tissues. The Navy has developed standard decompression tables (see FM 20-11-1) which are used to determine the rate of ascent and time required to stop for decompression. These tables must be followed during ascent to ensure the diver receives adequate decompression.

DECOMPRESSION DIVING

In a decompression dive, the body absorbs sufficient amounts of nitrogen to require controlled stops during ascent. These stops allow time for the body to off-gas the residual nitrogen. If the diver acsends too quickly or fails to make a scheduled decompression stop, the excess nitrogen will form bubbles. These bubbles may come out of the tissues and become lodged in joint areas, the spinal cord, or other places within the body. The lodged bubbles may cause some form of decompression sickness by blocking blood circulation or pinching nerves. Decompression sickness may range from slight pain to extensive paralysis; severe cases may result in complete stoppage of major organ functions.

Decompression dives must be performed using surface-supplied diving equipment. This equipment provides a continuous supply of air to the diver and communication between the surface team and the diver. A recompression chamber must be available at the site during decompression dives. Deep dives (dives over 100 feet salt water (FSW)) require the expertise of a master diver. Dives to 170 FSW require that a diving medical officer (DMO) be on call. Dives deeper than 170 FSW require a DMO be present to provide medical assistance (per Army Regulation (AR) 611-75). Finally, it is essential that divers performing decompression dives be in good physical condition and get at least eight hours of rest prior to the dive.

Divers are limited to the number of dives they can safely perform in a 24-hour period. Standard air-decompression tables in FM 20-11-1 are used to determine the amount of residual nitrogen in the tissues following a dive. The depth and time spent underwater will determine the amount of time a diver must remain on the surface before diving again.

Decompression dives place inordinate amounts of pressure on the body and require careful planning and specialized equipment. The diving team must be thoroughly trained, pay additional attention to safety details, and be prepared to respond to emergencies.

NO-DECOMPRESSION DIVING

No-decompression diving tables in FM 20-11-1 limit the maximum time a diver can spend at a specified depth without requiring decompression stops during ascent. Safe ascent can be made directly to the surface, at a prescribed rate, with no decompression stops. No-decompression dives can be performed in scuba or surface-supplied diving equipment.

MANNING

Diving operations require from 5 to 20 personnel. For instance, a mission requiring only one diver wearing scuba equipment and performing underwater work needs four additional divers for support from the surface. A mission requiring a LW team with one diver working at a depth of 50 FSW requires a 10-man crew; whereas, a mission requiring two divers working at a depth of 185 FSW requires a 20-man crew. Manning requirements depend on the mission, diving mode, and environment. Engineer diving teams are structured to work independently because the availability of outside diving support is limited. All assigned divers are required to support diving station functions such as operating the recompression chamber, handling hoses, and operating winches and air compressors. Additionally, engineer diving teams must provide their own drivers, mechanics, boat operators, medics, and radio operators. For many underwater engineer construction and salvage missions, two divers are normally required to perform the underwater work. Safety is a key consideration for manning requirements. If a diving team cannot be manned to safely operate in a hazardous work environment, mission accomplishment and diver safety may be compromised. Minimum staffing levels required for various types of diving operations are found in Appendix C of this manual and in AR 611-75.

MODES OF DIVING

Engineer divers use three distinct modes of diving.

• Scuba. Scuba operations are normally conducted to give the diver greater mobility to cover a larger area. The main advantages of scuba operations are quick deployment, mobility, depth flexibility and control, portability, and minimal surface-support requirements. Scuba is limited by time permitted at depth, lack of verbal communications, limited environmental protection, and remoteness from surface assistance. Scuba is used in water depths to 130 feet for underwater survey, inspection of potential work sites, searches, light work, and equipment and victim recovery. A scuba mission requires at least five personnel: one diver, one standby diver, one diving supervisor, two tenders, and one timekeeper/recorder. (The supervisor can perform as timekeeper.)

• Lightweight. Lightweight divers have unlimited air supplied by a flexible hose from the surface, good horizontal mobility, and voice and line-pull communications capabilities. Disadvantages include limited physical protection, limited vertical mobility, and the requirement for a large support platform. LW divers can be deployed to water depths of 190 feet for searches, inspections, light salvage, major ship repair, and working in enclosed spaces. LW missions require at least ten personnel: one diver, one standby diver, four tenders, one diving supervisor, one timekeeper/recorder, one air-control operator, and one communications operator.

• Deep-sea. Deep-sea divers have maximum physical and thermal protection, unlimited air supplied by a flexible hose from the surface, and voice and line-pull communications capabilities. A distinct disadvantage is that a large surface crew and a support platform are needed to operate air-support stations and recompression chambers during deep-sea diving missions. Deep-sea divers can be deployed to water depths of 190 feet for heavy salvage/repair and underwater construction. Deep-sea missions require at least 19 personnel: two divers, one standby diver, six tenders, one diving supervisor, one timekeeper/recorder, one air control operator, one communications operator, one master diver, and one diving officer. Four additional divers are required to operate air-support stations and recompression chamber.

EQUIPMENT

Specific equipment is required to protect and support the diving team. Diving operations can be conducted from the shore, piers, or floating platforms.

RECOMPRESSION CHAMBER

> **WARNING**
>
> **A recompression chamber must be located at the dive site if the diver does not have free access to the surface, enters an enclosed space underwater, or plans a decompression dive.**

A recompression chamber is a steel or aluminum cylinder large enough to accommodate a diver and necessary medical support personnel. The chamber may be used to treat diving injuries such as decompression sickness or arterial-gas embolisms. When pressurized with air, the chamber can simulate the pressure placed on the human body by a corresponding depth of water. Repressurizing the stricken diver in the chamber reduces the size of the lodged air bubbles. The stricken diver breathes 100-percent oxygen, which further aids in bubble resolution. Tables in FM 20-11-1 dictate times and depths for treatment of diving injuries. The chamber can also be used to perform surface decompression for certain types of decompression dives.

DIVE PLATFORM

If surface-supplied operations are conducted afloat, a suitable diving platform must be available for support. It should have certain minimum characteristics:

• Be able to safely carry all required equipment, including the recompression chamber.

• Provide adequate shelter and working area for support crew and divers, including a wash-down station and a warming area in cold weather.

• Be equipped with adequate navigation, signaling, and mooring gear.

• Include required lifesaving and safety gear.

• Be able to carry an additional small boat (if required) to rescue distressed divers, retrieve floating objects, and provide transportation in the event of an emergency or injury.

MEDICAL SUPPORT

All members of a diving team are trained to recognize diving-related disorders. Many types of diving disorders are life-threatening and require immediate treatment in a recompression chamber. Although the diving officer is in charge of the overall treatment of diving injuries, the master diver is the recognized authority and is responsible for the technical aspects of treatment. Each theater has a trained DMO to perform routine diving physicals and provide assistance for severe diving maladies. Each diving team must have an ETNCO immediately available to provide medical support. The ETNCO is trained in routine medicine as a medical noncommissioned officer (NCO) (military occupational speciality (MOS) 91B) and has received additional training in hyperbaric diving medicine. The hyperbaric medical training qualifies the ETNCO to diagnose and recommend treatment for divers stricken with diving maladies which require recompression therapy. He administers drugs as prescribed by the DMO.

SAFETY

All diving operations center around safety. Diving doctrine is driven by safety considerations. Safety dictates that divers be surface-tended when limited visibility or other hazardous environmental conditions warrant. A standby diver must be ready to deploy during all diving operations.

EXPLOSIVE ORDNANCE DISPOSAL

Engineer divers receive training in demolitions similar to that of MOS 12B. Additional training includes underwater demolition operations, use, hazards, and safety. Divers remove underwater mines and munitions by sympathetic detonation in the same manner as if the mines were surface laid. They can emplace the required explosives underwater, next to existing explosives, to detonate sympathetically. If the munitions must be disposed of or disarmed in another way, qualified underwater EOD units must be requested from the Navy. Engineer divers are neither trained nor equipped for EOD.

WATER-SAFETY MISSIONS

Engineer divers are not trained, qualified, or equipped to perform as certified lifeguards and should not be used as such. Special training and equipment are required to safely perform lifeguard responsibilities. Agencies such as the American Red Cross provide the necessary training and qualifications required for lifeguards.

Engineer divers can perform underwater recovery operations, but not within the time limits needed for emergency rescue. Diving skills are not recognized as a substitute for lifesaving skills. Tactical situations may require the use of engineer divers to prevent drowning. Such situations might include river-crossing operations where the far shore has been secured, or during amphibious operations in the COMMZ or other secured beachheads. Divers assigned to the Special Operations Forces are trained to operate in unsecured areas. The commander must carefully weigh the benefits of using engineer divers in water-safety missions because the loss of divers from their primary mission could have an adverse impact in the TO.

FLYING AFTER DIVING

Divers should not fly for at least 12 hours following a decompression dive or for two hours after surfacing from a no-decompression dive. If aircraft cabin pressure remains below an altitude of 2,300 feet, flying may be done after any type of air dive.

APPENDIX A

ENGINEER DIVING FORCE COMPOSITION
TOE 05-530LA00 - C&S DIVING DETACHMENT

PERSONNEL

Job Title	MOS	Rank	Quantity
Detachment Commander	21B5V	CPT	1
Operations Officer/Executive Officer	21B5V	1LT	1
Senior Diving Supervisor/Master Diver	00B50	MSG	1
Diver	00B30	SSG	2
Supply Sergeant	76Y30	SSG	1
Diver	00B20	SGT	2
PLL Clerk	76C20	SGT	1
ETNCO	91B20	SGT	1
Diver	00B10	SPC	3
TOTAL			13

EQUIPMENT

Line Number	Description	Quantity
B83924	Boat, Landing, Inflatable: Cotton Cloth, 7-Man	1
D89675	Chamber Recompression, Divers: 100 psi	1
E69790	Compressor Unit RCP, Air Ret, Gas and Diesel Driven, 88.5 cfm, 200 psi	4
F91490	Demolition Set Explosive: Electric and Semi-Electric	1
D32732	Diving Equipment Set: Open Circuit Scuba	11
D49154	Diving Equipment Set: Individual Swimmer Support Scuba	11
D32791	Diving Equipment Set: Photographic Support	1
D32859	Diving Equipment Set: Scuba Diving Support, Type A	1
D32927	Diving Equipment Set: Scuba Diving Support, Type B	1
G32815	Diving Equipment Set: Deep Sea	1
J35813	Generator Set Diesel Engine: 5kw 60Hz, l-3ph AC 120/208 120/240v Tactical Utility	1
L63994	Light Set, General, Illumination: 25 Outlet	1
N34334	Outboard Motor, Gasoline = 25-40 BHP	1
P92167	Pump Centrf Gas Drvn, Frame Mtd 2-inch, 170 GPM, 50 ft hd	2
P94290	Pump Centrf Gas Drvn, Wheel Mtd 4-inch, 600 GPM, 50 ft hd	2
P94359	Pump Centrf Gas Drvn, Wheel Mtd 6-inch, 1500 GPM, 60 ft hd	1
R44659	Radio Set: A.NPRC-87	1
W67980	Shop Equipment Cutting and Welding: Underwater Electric Less Power	1
S01373	Speech Security Equipment: TSEC/KY-57	1
W95537	Trailer Cargo: 3/4-ton, 2-wheel W/E	1
W95811	Trailer Cargo: 1 1/2-ton, 2-wheel W/E	1
T59346	Truck Cargo, Tactical: 5/4-ton 4X4 w/Commo Kit	1
T59482	Truck Cargo, Tactical: 5/4-ton 4X4 W/E M1OO8	1
X40146	Truck Cargo: 2 1/2-ton 6X6 w/winch W/E	1

APPENDIX B

ENGINEER DIVING FORCE COMPOSITION
TOE 05-530LC00 - LW DIVING TEAM

PERSONNEL

Job Title	MOS	Rank	Quantity
Team Leader	21B5V	1LT	1
Diving Supervisor/Master Diver	00B40	SFC	1
Diver	00B30	SSG	4
Diver	00B20	SGT	4
ETNCO	91B20	SGT	1
Diver	00B10	SPC	6
TOTAL			17

EQUIPMENT

Line Number	Description	Quantity
B83856	Boat, Landing, Inflatable: Assault Craft, Nylon Cloth, 15-Man	1
B83924	Boat, Landing, Inflatable: Cotton Cloth, 7-Man	2
D89675	Chamber Recompression, Divers: 100 psi	1
E69790	Compressor Unit RCP, Air Ret, Gass and Diesel Driven, 88.5 cfm, 200 psi	2
F91490	Demolition Set Explosive: Initiating Electric and Semi-Electric	2
D32723	Diving Equipment Set: Open Circuit Scuba	17
D49154	Diving Equipment Set: Individual Swimmer Support Scuba	17
D32791	Diving Equipment Set: Photographic Support	2
D32859	Diving Equipment Set: Scuba Diving Support, Type A	2
D32927	Diving Equipment Set: Scuba Diving Support, Type B	2
G32678	Diving Equipment Set: Lightweight	1
J35813	Generator Set Diesel Engine: 5kw 60Hz 1-3 ph AC 120/208 120/240v Tactical Utility	1
L63994	Light Set, General, Illumination: 25 Outlet	1
N34334	Outboard Motor, Gasoline = 25-40 BHP	2
R44657	Radio Set: AN/VRC-64	1
W67980	Shop Equipment Cutting and Welding: Underwater Electric Less Power	2
S01373	Speech Security Equipment: TSEC/KY-57	1
W95537	Trailer Cargo: 3/4-ton, 2-wheel W/E	1
W95811	Trailer Cargo: 1 1/2-ton, 2-wheel W/E	2
T59346	Truck Cargo: Tactical, 5/4-ton 4X4 w/Commo Kit	1
T59482	Truck Cargo: Tactical, 5/4-Ton 4X4 W/E M1OO8	2
X40009	Truck Cargo: 2 1/2-ton 6X6 W/E	2

APPENDIX C

MINIMUM STAFFING LEVELS FOR VARIOUS TYPES OF AIR DIVING

Diving Team	Depth of Dive (ft)[1]	Number of Divers[2]	A	B	C	D	E	F	G	H	I	J	K	L
1. Combat	0-130	one			1	1	1	1						4
		two			1	1	2	1						5
2. Scuba	0-130	one				1	1	1	1	1				5
		two				1	2	1	2[4]	1				7
3. Surface-Supplied, Lightweight	0-100	one				1	1	1	4[5]	1			1	1	10
		two				1	2	1	6[5]	1	1	1	1	13
	100-170	one	1	1		1	1	1	4	1	4[6]	1	1	16
		two	1	1		1	2	1	6	1	4[6]	1	1	19
	below 170	one	1	1	1[3]	1	1	1	4	1	4	1	1	17
		two	1	1	1[3]	1	2	1	6	1	4		1	20
4. Surface-Supplied, Deep Sea	0-100	one				1	1	1	4	1			1	1	10
		two				1	2	1	6	1			1	1	13
	100-170	one	1	1		1	1	1	4	1	4[6]	1	1	16
		two	1	1		1	2	1	6	1	4[6]	1	1	19
	below 170	one	1	1	1[3]	1	1	1	4	1	4	1	1	17
		two	1	1	1[3]	1	2	1	6	1	4	1	1	20

Notes:

1. A hyperbaric chamber is required on all dive sites during any planned or anticipated decommission dives and during diving operations where free access to the surface is restricted.
2. See FM 20-11-1 for air diving operations requiring more than two divers.
3. A diving medical officer is required to be on call for all planned or anticipated decompression dives. He must be present for all dives deeper than 170 feet or when particularly hazardous diving operations are being conducted.
4. One tender per diver when divers are surface-tended. If using buddy system, one tender required for each buddy pair.
5. For dives 0-60 feet, only one tender per diver is required when using lightweight equipment.
6. If Note 1 does not apply, chamber crew is not required.
7. Key for table:
   A. Diving officer.
   B. Master diver.
   C. Diving medical officer (for chamber divers diving under no decompression limits, a trained diving medical technician may be substituted).
   D. Diving supervisor. The diving supervisor for scuba (diving teams 2, 3, and 4) is either a master diver (MOS 00B40/50) or a first-class diver (MOS 00B) in organizations and activities authorized these positions by TOE or TDA. The diving supervisor for diving teams 3 and 4 (100 ft or below) is a master diver (MOS 00B).
   E. Diver.
   F. Standby diver.
   G. Tender.
   H. Timekeeper/recorder.
   I. Chamber crew. An ETNCO (Diving) should be the insider tender during hyperbaric chamber treatments.
   J. Air control console operator.
   K. Communications operator.
   L. Total personnel required. The total required does not include safety boat crew or personnel required to operate special equipment and tools.

REFERENCES

SOURCES USED
These are the sources quoted or paraphrased in this publication.

AR 611-75. *Selection, Qualification, Rating and Disrating of Marine Divers.* 16 August 1985.
AR 611-201. *Enlisted Career Management Fields and Military Occupational Specialties.* 1 October 1990.
FM 5-480. *Port Construction and Repair.* 12 December 1990.
FM 20-11-1. *Military Diving (Volume 1).* 20 September 1990.
FM 63-4. *Combat Service Support Operations - Theater Army Area Command.* 24 September 1984.
FM 100-16. *Support Operations: Echelons Above Corps.* 16 April 1986.
NAVFAC P-990. *Conventional Underwater Construction and Repair Techniques.* Undated.
SO4OO-AA-SAF-O1O. *US. Navy Salvage Safety Manual.* December 1988.

DOCUMENTS NEEDED
These documents must be available to the intended users of this publication.

DA Form 2028. *Recommended Changes to Publications and Blank Forms.* February 1974.

READINGS RECOMMENDED
These readings contain relevant supplemental information.

CW 01333. *Civil Works Specifications, U.S. Army Corps of Engineers, Hydrographic Surveying Services.* November 1990.
EM 1110-2-1003. *U.S. Army Corps of Engineers, Engineering and Design: Hydrographic Surveying.* February 1991.
FM 3-4. *NBC Protection.* 21 October 1985.
FM 3-5. *NBC Decontamination.* 24 June 1985.
FM 3-10-1. *(SRD) Chemical Weapons Employment.* 8 April 1988.
FM 5-25. *Explosives and Demolitions.* 10 March 1986.
FM 5-36. *Route Reconnaissance and Classification.* 10 May 1985.
FM 5-100. *Engineer Combat Operations.* 22 November 1988.
FM 5-103. *Survivability.* 10 June 1985.
FM 5-104. *General Engineering.* 12 November 1986.
FM 5-105. *Topographic Operations.* 9 September 1987.
FM 5-116. *Engineer Operations: Echelons Above Corps.* 7 March 1989.
FM 5-134. *Pile Construction.* 18 April 1985.
FM 5-233. *Construction Surveying.* 4 January 1985.
FM 5-446. *Military Nonstandard Fixed Bridging.* 3 June 1991.
FM 8-8. *Medical Support in Joint Operations.* 1 June 1972.
FM 8-10. *Health Service Support in a Theater of Operations.* 2 October 1978.
FM 8-21. *Health Services Support in a Communications Zone.* 1 November 1984.
FM 9-6. *Munitions Support in Theater of Operations.* 1 September 1989.
FM 10-60. *Subsistence Supply and Management in Theaters of Operations.* 29 December 1980.
FM 10-67. *Petroleum Supply in Theaters of Operations.* 16 February 1983.
FM 14-6. *Comptroller/Finance Services in Theaters of Operations (Keystone).* 10 September 1981.
FM 19-30. *Physical Security.* 1 March 1979.
FM 19-40. *Enemy Prisoners of War, Civilian Internees and Detained Persons.* 27 February 1976.
FM 20-11-2. *Military Diving (Volume 2).* August 1991.
FM 20-22. *Vehicle Recovery Operations.* 18 September 1990.
FM 20-32. *Mine/Countermine Operations.* 9 December 1985.
FM 31-11. *Doctrine for Amphibious Operations.* 1 August 1967.
FM 31-12. *Army Forces in Amphibious Operations (The Army Landing Forces).* 28 March 1961.
FM 31-25. *Special Forces Waterborne Operations.* 30 September 1982.
FM 31-82. *Base Development.* 3 June 1971.
FM 41-10. *Civil Affairs Operations.* 17 December 1985.
FM 55-1. *Army Transportation Services in a Theater of Operations.* 30 November 1984.

FM 55-60. *Army Terminal Operations.* 18 May 1987.
FM 55-509. *Marine Enginemans Handbook.* 3 October 1986.
FM 90-13 (HTF). *River Crossing Operations (How to Fight).* 1 November 1978.
FM 101-5. *Staff Organization and Operations.* 25 May 1984.
FM 101-5-1. *Operational Terms and Symbols.* 21 October 1985.
FM 101-10-1/1 *Staff Officers Field Manual - Organizational, Technical and Logistical Data (Volume 1).* 7 October 1987.
FM 101-10-U2. *Staff Officers Field Manual - Organizational, Technical, and Logistical Data, Planning Factors (Volume 2).* 7 October 1987.
TM 5-343. *Military Petroleum Pipeline Systems.* February 1969.
TM 5-622. *Maintenance of Waterfront Facilities.* June 1978.

GLOSSARY

1LT   first lieutenant

AC   alternating current

AR   Army regulation

area engineer engineer representative from the engineer district or engineer division responsible for contracting construction projects with civilian contractors and monitoring their progress.

Army water terminal Army-controlled harbor or port facilities.

arterial gas embolism a pulmonary barotrauma caused by the expansion of gas within the lungs. Usually as a result of air breathed under pressure and not exhaled during ascent. Gas could have become trapped by mucus obstruction resulting from lung congestion or by diver reacting with panic to a difficult situation and holding his breath during ascent without realizing it. When the gas expands sufficiently, the pressure will force the gas through the alveolar walls into the surrounding tissues and into the bloodstream. Divers with emboli blocking blood flow to the major organs show definite symptoms normally within only a few minutes. The diver must be diagnosed quickly and correctly. Immediate recompression within the chamber is required to prevent permanent damage or loss of life.

attached an attached engineer element is commanded by its supported unit, maintains liaison and communications with supported unit, is task organized by the supported unit, responds to support requests from its supported unit, has its work priorities established by the supported unit, has its spare work available to its supported unit, requests support from its supported unit, and receives logistical support from its supported unit. When attached, the engineer element is provided administrative and logistical support. However, some special logistical and administrative needs are still provided by the parent unit.

ASG   area support group

bearing pile a long, slender column usually of timber, steel, or reinforced concrete driven into the ground to carry a vertical load.

BHP   brake horsepower

buoy float anchored to mark objects or locations under water.

C&S   control and support

centrf   centrifugal

cfm   cubic feet per minute

combat zone that area required by combat forces for conducting operations, usually forward of the Army rear boundary.

commo   communication

communications zone rear part of the TO behind the CZ that contains the LOC and supply supporting combat forces.

COMMZ   communications zone

CPT	captain
CW	civil works
CZ	combat zone
DA	Department of the Army

dewater to remove water.

direct support an engineer element in a direct support role is commanded by its parent unit, maintains liaison and communications with supported and parent units, may be task-organized by its parent unit, provides dedicated support to a particular unit, responds to support requests from its supported unit, has its work priority established by the supported unit, has its spare work effort available to its parent unit, requests additional support from its parent unit, and receives logistical support from its parent unit.

diver tender surface member who assists the diver and standby diver during equipment checks and dressing and undressing into equipment. He tends the umbilical lifeline and reports to the diving supervisor. Two tenders are required for each surface-supplied diver in water depths of 60 feet of more and for the standby diver. One tender is required for each scuba diver. The tender should be a qualified diver.

DMO	diving medical officer

dolphin system a cluster of closely driven piles used as a fender for a dock as a mooring or guide for boats.

drvn	driven

dry dock an enclosed dock that can be dewatered to provide a stable, dry platform for use during the repair of ships.

EAC	echelons above corps
EM	engineer manual
ENCOM	engineer command
EOD	explosive ordnance disposal
ETNCO	emergency treatment noncommissioned officer
F	Fahrenheit
FEBA	forward edge of the battle area

fender system a system of wood or rubber devices designed to absorb the shock of ship movement and protect the pier structure.

flattening removal of superstructure and crushing hull with demolition into the port bottom.

FLOT	forward line of own troops
FM	field manual
fps	feet per second
FSW	feet salt water

ft	foot, feet
GPM	gallons per minute
hd	head

high watermark the highest point on shore that water reaches during high tide.

HQ	headquarters

hull the lowermost, watertight portion of a vessel that gives it buoyancy.

Hz	hertz
IAW	in accordance with
IDSM	intermediate direct support maintenance
IGSM	intermediate general support maintenance
kw	kilowatt
LIC	low-intensity conflict
LOC	lines of communication
LOTS	logistics over the shore operations

low watermark lowest point on shore that is exposed during low tide.

LW	lightweight

marine railway a rail system extending below water designed to bring harbor craft out of the water for repair.

min	minute

mooring site an area designated for the temporary anchorage of vessels. The site is provided with mooring buoys and designed to allow sufficient space for vessels swinging on a moor.

MOS	military occupational specialty
MSG	master sergeant
mtd	mounted
NAVFAC	Naval Facilities Engineering Command
NBC	nuclear, biological, chemical
NCO	noncommissioned officer
OPDS	offshore petroleum distribution system
ph	phase

pier a structure extending into navigable waters used as a landing and for the loading and unloading of vessels.

PLL	prescribed load list

psi   pounds per square inch

RCP   reciprocating

rec   receiver

recompression chamber apparatus which is pressurized with air to decompress a diver or treat a pressure-related diving illness after surfacing.

RWCM   regional wartime construction manager

SALMS   single anchor leg mooring system

scuba (self-contained underwater breathing apparatus) an apparatus used for breathing while underwater.

SFC   sergeant first class

SGT   sergeant

ship channel the deeper part of a harbor, river, or strait designated, marked, and maintained to permit the safe passage of ships.

ship husbandry work performed on vessels for repair or maintenance.

SPC   specialist

SRC   survival recovery center

SSG   staff sergeant

surface-supplied air diving equipment where the breathing air is supplied through flexible rubber hoses to the diver from compressors or storage facilities on the water's surface.

TA   theater Army

TAACOM   Theater Army Area Command

TAHQ   theater Army headquarters

temp   temperature

timekeeper/recorder records each diver's descent events and time and bottom time. Calculates decompression obligation requirements. Completes dive summary records and official transcript of dive.

TO   theater of operations

TOE   table(s) of organization and equipment

topo   topography

US   United States (of America)

USAES   United States Army Engineer School

v   voltage

w/   with

W/E with equipment

wharf a structure built along, or at an angle from, the shore of navigable waters so that ships may lie alongside to receive and discharge cargo.